Antonio Alberto Vela Avila
Julio Cesar Canul Ek
Jose Lazcano Pacheco

ENERGY DIAGNOSIS

Antonio Alberto Vela Avila
Julio Cesar Canul Ek
Jose Lazcano Pacheco

ENERGY DIAGNOSIS

THEORY AND REALIZATION OF AN ENERGY DIAGNOSIS

ScienciaScripts

Imprint

Any brand names and product names mentioned in this book are subject to trademark, brand or patent protection and are trademarks or registered trademarks of their respective holders. The use of brand names, product names, common names, trade names, product descriptions etc. even without a particular marking in this work is in no way to be construed to mean that such names may be regarded as unrestricted in respect of trademark and brand protection legislation and could thus be used by anyone.

Cover image: www.ingimage.com

This book is a translation from the original published under ISBN 978-620-2-11469-1.

Publisher:
Sciencia Scripts
is a trademark of
Dodo Books Indian Ocean Ltd. and OmniScriptum S.R.L publishing group

120 High Road, East Finchley, London, N2 9ED, United Kingdom
Str. Armeneasca 28/1, office 1, Chisinau MD-2012, Republic of Moldova, Europe
Printed at: see last page
ISBN: 978-620-5-73649-4

ACKNOWLEDGMENTS

The authors would like to thank the **Tecnológico Nacional de México** for giving us the opportunity to train competitive professionals in science, technology and other areas of knowledge, committed to the economic, social and cultural development and sustainability of the country. Likewise, we thank the **Tecnológico de Campeche** for helping us to materialize our efforts in the production of quality human capital.

Technology in the province enlarges the nation

SUMMARY

This document represents the work carried out at INERCI in the state of Campeche, Mexico, where energy diagnostics were performed in the years prior to this publication.Here we show the basic concepts of the types of energy diagnosis, the procedures and methodology as well as the results obtained from the analysis of a maquiladora company. In which data and parameters were analyzed and suggestions are also proposed to the company.

CONTENT

INTRODUCTION

The constant technological improvement of electrical energy measurement equipment requires us to know the improvements of our work tools in order to provide a better service to our customers.Electricity is one of the most widely used energy sources in Mexican households. It is mainly used for air conditioning, refrigeration, lighting and various household appliances. Every two months, the electricity bill has a strong impact on our pockets.As it is intended to obtain the reduction of consumption, through the measurement of energy of the equipment installed in the study area, the indexes that allow to make an energy diagnosis and based on this determine both the current state of consumption with respect to the nominal load, as well as to determine possible sources of energy savings, determining the anomalies presented by the system, to formulate measures in the short, medium and long term.In general terms, the efficient use of electrical energy means better production results with fewer resources, which will translate into lower manufacturing costs, more products with less waste and lower energy consumption.Business customers require advice on the metering of their utilities, so energy diagnostics are no longer limited to energy saving advice, but also include billing and therefore metering information.

JUSTIFICATION

When we want to determine the saving possibilities of an electrical installation, a diagnosis is necessary.According to Neagu Bratu and Eduardo Campero (1995), diagnosis means an inspection to determine the physical condition of a facility and to provide information on the efficiency with which electrical energy is transported and converted. At the same time, it should help determine better operating options.The diagnosis must ensure that in an electrical installation, electrical energy is distributed to the connected equipment in a safe and efficient way, one of the biggest problems in industrial and home installations is the loss of energy. This is reflected in the economy.By providing adequate information, it helps to better manage energy, this means that the necessary energy will be consumed, in the sense that it represents a lower cost, however, this should not affect safety, production volume or the quality of the product or service. Saving energy means reducing energy consumption while achieving the same results obtained with more electrical energy.

OBJECTIVES

- Verify that the installation is safe and does not represent a risk for users or for the equipment it supplies or that is nearby.
- Identify problems that cause low efficiency in order to avoid unnecessary consumption.
- Elaborate technical measures according to the data obtained from a previous inspection.
- Implement the technical measures elaborated, which help to manage energy in an adequate manner, without affecting the volume of production or the quality of the product or service for which it is designated.

GENERAL

From the company.
Company information.
Company name: INERCI. **Address.** KNOWN **Telephone:** 981 688 5543

Background.

Customer service is a priority for the company, which is why it uses technology to be more efficient, and continues to expand its service, taking advantage of the best technologies to provide service even in remote areas and dispersed communities.

INERCI is a company dedicated to electrical installations to local companies, and maintains integrated all the processes of the electrical service.

Mission.

To provide the public electric energy service with criteria of sufficiency, competitiveness and sustainability, committed to customer satisfaction, the development of the country and the preservation of the environment.

Vision.

VISION TO 2030

To be an energy company, one of the best in the electricity sector worldwide, with international presence, financial strength and additional income from services related to its intellectual capital and physical and commercial infrastructure.A company recognized for its customer service, competitiveness, transparency, service quality, staff capacity, technological vanguard and application of sustainable development criteria.

CHAPTER I
DESCRIPTION AND ANALYSIS OF AN ENERGY DIAGNOSIS

1.1. LEVELS OF STUDY

An energy diagnosis is the application of a set of techniques to determine the degree of efficiency with which energy is used. In order to determine areas of opportunity and potential savings by type of measure, evaluating its profitability. There are 3 levels of diagnostics detailed below.

1.1.1. Energy diagnosis 1st level.

Its main objective is to obtain an overall balance of energy and potential savings that do not require investment, focuses on training users to have a change of habit that leads to energy savings, you can get to make a control of lights using them only if necessary, engine shutdowns when working in a vacuum without any benefit, control of air conditioners for scheduled use, etc..

This level consists of:
1. Information gathering.

- Energy consumption.

- Production levels

- Main equipment.

2. Tour of the facilities, visual inspection and review of consumer equipment.
3. Determine energy saving potentials.
4. Provide recommendations:

- Waste disposal.

- Proper operation of equipment and systems.

- Change of personnel habits.

5. Projects with no or very low investment (1 to 6 months) are presented.
6. Implementation of cost-saving measures.

In this diagnosis, an in-depth analysis of energy use is not necessary; implementation measures should be proposed immediately.

1.1.2. Energetic diagnosis 2nd level.

At this level, specific energy balances should be obtained, as well as energy saving potentials with or without investment, which will be applied to the process.An energy efficiency assessment will be carried out in intensive areas and equipment involved.This process requires a detailed analysis of the consumption history where the operating equipment is involved, it is necessary to know the process that the company manages and what are its specific energy consumptions, these should not be assumed, for this reason it is necessary to have an accurate measurement with appropriate instruments.

This level consists of:
1. Analysis of energy aspects, in main and auxiliary equipment and systems.
- Consumption
- Production
- Billing
2. Electrical measurements
- Analyzer
- Multimeters
- Luxmeter
- Flow meter
3. Determination of energy savings potential.
- Energy balances
- Estimation of efficiencies
- Identification of new technologies.
4. Technical-economic analysis of the measures
5. Classify short- and medium-term projects (7 months to 3 years of recovery).
6. Selection of projects to be carried out:
- Cost-benefit ratio.
7. Implementation of cost-saving measures.

1.1.3. Energy diagnosis 3rd level.

At this level, a thorough analysis of the operating conditions and design basis of a facility is required, using specialized measurement and control equipment.It requires complete information on the flow of materials, fuels, electrical energy, as well as pressure, temperature and the properties of the different substances and/or currents with which the process works.The technical measures generated in this process are usually medium to long term, involving modifications to equipment, processes and the technologies used.

This level consists of:
1. Exhaustive process analysis.
- Change in operating habits.
- Elimination or change of shifts.
2. Identification of state-of-the-art technologies that will replace the previous one.
3. Technical feasibility and economic profitability analysis.
4. Projects with high investment levels (payback period of 3 years or more) are presented.
5. Presentation of the projects to management.
6. Authorization and mechanics of financing
7. Project start-up.

1.2. ANALYSIS OF ELECTRICITY BILLS AND INTERPRETATION OF TARIFFS.

Electricity tariffs are the specific provisions containing the fees and conditions governing electricity supplies grouped into each class of service.

1.2.1. Tariff electricity

Rates are officially identified by their number and/or letter(s) and only in those cases where it is necessary to supplement the denomination.

Table 1 Description of the different rates

IDENTIFICATION	TITLE
01	Domestic service
1ª	Domestic service for locations with minimum summer temperatures of 25 degrees centigrade
1B	Domestic service for locations with minimum summer temperature of 28 degrees centigrade
1C	Domestic service for locations with minimum summer temperature of 30 degrees centigrade
1D	Domestic service for locations with minimum summer temperature of 31 degrees centigrade
1E	Domestic service for locations with minimum summer temperature of 32 degrees centigrade
02	General service up to 25 KW of demand
03	General service for more than 25 KW of request
05 and 05th	Service for street lighting
06	Drinking water pumping service or black public service
07	Temporary service
09 and 09M	Water pumping service for irrigation purposes agricultural
O-M	Ordinary rate for general service in medium voltage with demand over 10 KW and under 100 KW
H-M	Hourly rate for general service in medium voltage, with demand of 100 KW or more
H-S	Hourly rate for general high service voltage, sub-transmission level
H-T	Hourly rate for general high service voltage, transmission level
H-SL	Hourly rate for general high service tension subtransmission level, for long term use
I-15 E I-30	Interruptible service rates
R	Hourly rate for backup service for medium voltage and high voltage failure and maintenance (HM-RM, HS-RM, HT-RM)
DAC	Tariff from service domestic from high consumption

1.2.2. Classification of the rates.

According to their application, tariffs are classified as follows:

1.2.1. Specific

The specific tariffs are those that apply to electric energy supplies used for the purposes indicated therein. The following tariffs correspond to this group: 1, 1a, 1b, 1c, 1d, 1e, 5, 5a, 6, 9 and 9m.

1.2.2. General

the rates for general uses are those applicable to any electric service, except for the specific ones mentioned above; the following rates correspond to this group: 2, 3, 7, o-m, h-m, h-m, h-s, h-t, h,sl, h-tl, i-15 and i30.

1.2.3. Back-up

These are the tariffs for medium and high voltage backup service for individuals that use the electric power generation modalities and establish the backup options for failure and maintenance backup, failure and backup for programmed maintenance, to this group belong: HM-RF, HS-RF, HM-RF, HS- RM.

1.2.3. Charge to hire

It is the sum of the power of the equipment, appliances and devices that the customer will connect to its installations, expressing the total value in KW and that the applicant will state when requesting the supply.

1.2.3.1. Rules For Determining The Charge To Be Contracted

A. The application for the supply of electric energy must establish the load to be contracted, which is established based on the data provided by the applicant, serving to determine the feasibility of supplying the service and selecting the necessary metering equipment.

B. In the event of standby equipment, i.e., equipment that cannot opérate simultaneously with the one it is intended to replace, shall not be counted as an integrated part of the load.

C. For incandescent lamps, the capacity in WATTS of each one of them shall be added.

D.Lamps that require a starting device, their rated capacity plus 25% (twenty-five percent) shall be taken to consider the capacity in WATTS of the auxiliary equipment required for their operation. This percentage may vary according to the results obtained by the supplier, at the customer's request, from capacity tests of the auxiliary equipment, in which case, the contract may be modified taking into account such results.
E.On X-ray machines, welding machines, spot welders, etc. It shall be taken
its nominal capacity in voltamperes at a power factor of 90%, i.e., to obtain the power in WATTS, multiply by 0.90 the capacity in voltamperes.
F.In the case of electric motors, the capacity of each one of them will be taken individually by applying the table in which the efficiency of the motors is contemplated. This table is shown below

1.2.4. Complaint by contracting.

It is the demand that the supplier and the customer initially agree upon in the respective contract, under rules established in the tariffs themselves that must be complied with under C.F.E.'s supervision in order to ensure that its value corresponds to the power requirements of the service and, if applicable, serve as a real element to determine the respective value of the guarantee deposit.

1.2.4.1. Modification of the contracted load and demand.

The supplier will be responsible for the supply under the conditions agreed upon, according to the contracted load limit. Variations in said load, caused by an increase in the customer's electric energy requirements, obliges the supplier to consult with CFE as to whether the energy can be supplied and under what conditions, both technical and economic. The customer will communicate in writing to the supplier the new load and demand; if applicable, within fifteen days following the date on which the contracted load and demand varied, and must adjust the amount of the guarantee deposit for the new supply conditions. It will be cause for suspension of the supply, in terms of article 26, section IV of the law (it states that the suspension of the supply of electric energy must be made when it is proven that the use of electric energy in conditions that violate what is established in the respective contract). Failure of the customer to timely notify increases in demand that exceed the connected load; if applicable, in the contracted load and demands, if such increases originate or could originate disruptions to the general public service.When on three successive occasions the maximum demand measured exceeds the contracted demand, the customer will be required to modify its contract and update the guarantee deposit, taking the average of the three as the new minimum contracted demand.Regardless of the update of the security deposit, the collection of contributions for specific works to support the new contracted load and/or the contribution per KVA in accordance with the tariff will be applicable.The amount of the guarantee deposit will be composed of the amount of the deposit of the contract being modified, plus the amount corresponding to the application of the deposit installments established in the tariffs in effect, to the difference between the demand established in the contract being modified and the contracted demand established when entering into a new contract.

1.2.4.2. Supply voltage.

These services are supplied at medium voltage, i.e. from 7600 volts to 34500 volts, as requested by the user.

1.2.4.3. Load and demand for contracting.

The load to be contracted will be the sum of the power in Kilowatts of the equipment, appliances and devices that the customer claims to have connected.The demand to be contracted will be initially set by the customer, its value will not be less than 60% of the total connected load, nor less than 100 KW of the capacity of the largest installed motor or appliance.In the event that 60% of the total connected load exceeds the capacity of the customer's substation it will only take the capacity of the customer's substation at a factor of 60%.Any fraction of Kilowatts will be taken as a full Kilowatt.

1.2.5. Other options tariffs.

The supplier is authorized to enter into an agreement with the customers of this tariff who so request to agree their billing under the contracted demand option.

1.2.5.1. Change from hm to om rate.

When the customer maintains for 6 consecutive months, both a maximum demand measured in peak, intermediate and base periods of less than 100 kW, it may request its incorporation to the supplier.

1.2.5.2. Basic billing

The basic billing is integrated by adding the charges for billable demand, the authorized fees for peak, intermediate and base consumption that are recorded in a normal billing period and according to the applicable tariff regions and schedules. Due to the importance of these supplies, the consumption period will be from 0:00 hours on the 1st day of the billing month to 24:00 hours on the last day, for which the meters are activated with a reading freeze function that will keep these values in memory, allowing readings to be taken every 1st day of each month.In months when there is a change of season on the first or last day of the month, the meter will freeze the reading at the change of season and the reading at the end of the month.

1.2.5.3. Low voltage measurement charge

In case the measurement is made on the secondary side of the customer's substation, a 2% charge will be applied to the basic billing.

1.3. THE POWER FACTOR

The ratio of real power to apparent power is a circuit is known as the power factor. Thus, it is a way of indicating what portion of the total current and voltage produces power. When the current and voltage are in phase, the power factor will be 1; the real power will be the same as the apparent power. When the current and voltage form 90°, the power factor will be 0. That said, the factor can vary between 0 and 1, this value is expressed as a percentage and ranges from 0% to 100%.

1.3.1. Power factor analysis

In order to understand why the power factor appears in electrical installations, an analysis is made of the different elements that constitute the load of an installation.

1.3.1.1. Capacitor

The capacitor or capacitor is the element capable of storing electric charge. Its behavior is as follows: if we connect a battery to the capacitor, the current will rise almost instantaneously up to a certain peak value, from where it descends exponentially to zero (remaining electrically charged). If an arrangement is achieved (adjustable resistance) that allows a gradual decrease in the battery voltage, the current will begin to flow in the opposite direction and reach a maximum negative value when the voltage is zero. If at a certain moment the voltage is increased in the opposite direction to the previous one, the current starts to return to zero again. That is, the current always goes one step ahead of the voltage, in summary can be understood in the capacitor current precedes the voltage.

1.3.1.2. Inductance

In an inductance, energy is stored in the form of a magnetic field. If we were to apply the voltage of a battery to an inductive element, the current would grow exponentially. This current establishes a field in the inductor core that opposes sudden changes. This causes there to be a delay in the flow of current through the coil. If we were able to decrease the battery voltage, it would be observed that the current would also decrease. However, at the moment when the voltage reaches zero, the current field would decrease. magnetic field of the inductor would oppose the current to be zero and induce a voltage that would cause it to continue flowing. It is understood that in an inductance the voltage precedes the current. If we consider an ideal capacitance or inductance connected to an alternating voltage source, the phase shift of the current with respect to the voltage will be 90° forward in the case of capacitance, and 90° backward in the case of inductance. Current and voltage, in any given circuit, can have an electrical phase shift between zero and 90° with current preceding voltage, or vice versa.

1.3.1.3. Active power

If a direct current of constant value is passed through a resistor (R), the electrical energy is transformed into thermal energy.

Joule's law states that heat energy is equal to power per unit time. heat energy = R - 1^2 - t = P - t
This power (P} is known as active power.

1.3.1.4. Reactive power.

In the case of a circuit with a purely capacitive or inductive element, the energy does not change form, it is only stored. In other words, the source delivers energy to the capacitive or inductive element, which stores it and in turn delivers it when the source is de-energized. If the circuit is connected to an alternating current source, the energy passes from the source to the capacitor (or inductor) in the first quarter cycle and returns to the source in the next quarter cycle. This energy associated with an ideal capacitor or an ideal inductor is called reactive. In the same way it is called reactive power "Q" to the capacitive or inductive power that multiplied by the unit of time produces this type of energy. It is called capacitive when the current precedes the voltage, and inductive when the voltage is higher than the voltage. precedes the current. For both cases (with ideal elements) there is an offset of 90° with respect to the active power.

1.3.1.5. Apparent power.

Electrical installations are a combination of resistive, inductive and capacitive elements, so the power required has an active and a reactive component. The vectorial sum of these two components is known as apparent power "S". This power is the one used to calculate the cross-sections of the conductors and the other elements of the installation.

1.3.1.6. Mechanical analogy.

Mechanical work is defined as the product of the increment of the distance by the force vector in the direction of displacement: force times distance. Mechanical (active) power is equal to the rate of work production, or, in other words, the number of units of work performed in a unit of time: force times velocity (distance times time).If the force is applied in a direction other than the direction of motion, it can be considered as the sum of two force vectors, located: one on the axis of the direction of motion, the other at 90° to the first. Of these components, only the first (along the motion) produces work. For this reason, when the angle of application of the force increases, the active component decreases and in turn the work produced. Therefore, if the same speed is to be maintained, the application of a greater force is required (the increase must be proportional to the cosine of the angle of application). A similar phenomenon occurs in an electrical network: the greater the angle between voltage (force applied in the analogy) and current (speed), the more apparent power is required to supply the same active power.

1.3.2. Definition of the power factor.

In electrical installations there are usually devices that transform energy into heat or work together with inductive and capacitive elements that do not do work. So there is practically always UR angle between voltage and current known as phase angle.The power factor is the quotient of the ratio of total watts to total RMS volt-amperes (root-mean-square, root-mean-square or root-mean-effective value), i.e., the ratio of active power to apparent power. When current and voltage are sine functions and ϕ is the angle of phase shift between them, the cosine of ϕ is the power factor (p.f.). Then, the p.f. depends on the phase shift between voltage and current, which in turn depends on the load connected to the circuit. the active power is equal to the product of the effective (RMS or quadratic) values of the voltage "V" and the current "I" times the cosine of the phase angle between them

P = V - 1 coscp
Where: ϕ is the phase angle between voltage and current.

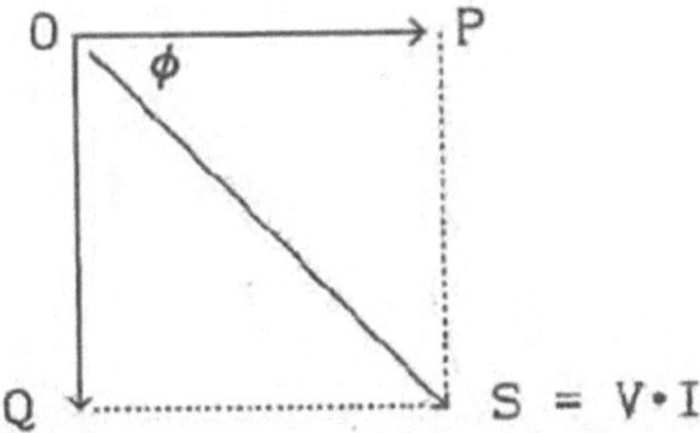

Figure 1 Power vector diagram

Therefore:

$$S = \sqrt{P^2 + Q^2}$$

The load of an installation is mainly constituted by electrical equipment (motors and transformers) made of coils (inductances). It is normal to find that the inductive load predominates over the capacitive one, that is to say, generally the current is lagging behind the voltage, so it is common to speak of lagging power factor.

1.3.3. Consequences of a low power factor

The current flowing in the conductors can be broken down mathematically (not physically) into two components: one coinciding with the active power and the other with the reactive power.
I.e.

$$f.\,p. = \cos\varphi = \frac{P}{S} = \frac{P}{\sqrt{P^2 + Q^2}}$$

Where:

1 = Total current

$1a$ = Current component (in phase with voltage)

1_r = Reactive component lagging 90° behind the voltage

The power factor decreases or increases according to the cosine function of the phase angle. If you have a load where the phase shift of the current (lagging) with respect to the voltage is very close to 90°, the p.f. will be very close to zero and the re active component of the current will be very large compared to the active component.

1.3.4. Power factor compensation.

Industrial electrical installations whose load is mainly composed of induction motors have a lagging power factor. For this reason it is necessary to compensate the inductive load with capacitive load.
The most economical and simplest solution is usually to install capacitor banks that provide the reactive kVA's necessary for theThe p.f. is higher than what is stipulated in the supply contract. In fact, the supply companies themselves use this system to compensate the f.p. of their transmission and distribution network.The power factor can also be compensated by using synchronous motors instead of induction motors, but once the necessary kilovar (kVA reactive) is defined, the problem requires an economic rather than a technical analysis. The amount of kvar needed to improve the fp. is obtained from the reactive power required by the equipment constituting the installation. In many occasions this is done with the measurement of the first month of operation of the equipment. Consider that the initial conditions are:

$$I = \sqrt{I_a^2 + I_r^2}$$

$$f.\, p. = \cos\varphi = \frac{I_a}{I} = \frac{I_a}{\sqrt{I_a^2 + I_r^2}}$$

And the desired conditions are:

It follows that the capacitor bank to be installed must supply a three-phase reactive power:

$$S_1 = \frac{P}{\cos\varphi_1} \qquad\qquad Q_1 = \sqrt{S_1^2 - P^2}$$

$$S_2 = \frac{P}{\cos\varphi_2} \qquad\qquad Q_2 = \sqrt{S_2^2 - P^2}$$

Where:

V = nominal voltage applied to the capacitor bank.

I = current to flow through the capacitor bank.

$$Q = \sqrt{3}\, V - I = Q_1 - Q_2$$

It may be necessary to consider stepped capacitance values to allow for normal load variations. It is also important to monitor voltage behavior, especially if there are many on and off switching operations close to the capacitors.

CHAPTER II
DIAGNOSTIC PROCEDURE ENERGETIC

2.1. INITIAL TOUR OF FACILITIES

Identification of:

- Type of technology installed
- Location of equipment
- Type of panels (control room)
- Physical condition of equipment
- Type of processes
- Work schedules
- Equipment on without use
- Forms and sources of energy used
- Energy substitution possibilities
- Possibilities of self-generation and cogeneration.

2.2 ELECTRICAL MEASUREMENTS FOR THE DIAGNOSTICS

The objective of the electrical measurement is to obtain the main electrical parameters that allow to know the electrical behavior of the company and/or of the equipment existing in the company.Knowing the electrical parameters of the company and/or equipment is of fundamental importance in order to proceed in the evaluation of energy saving potentials. The electrical measurement has the following purposes:

A. Verify equipment operation times and schedules.

B. Detect demand peaks.

C. Detect power factor anomalies.

D. Detect phase unbalance.

E. Have antecedents a the implementation of measures of saving measures energy saving measures and programs to control electricity demand.

F. Compare electric metering data with billing data.

Electrical measurements can be general or punctual depending on the company's objectives and resources. The reliability of the measurements depends directly on the sophistication of the measuring instruments used.

2.2.1. Measurements

These measurements are performed with a portable network analyzer or with a hook ammeter, multimeter and factor meter.Spot measurements shall be taken under normal operating conditions of the equipment and shall be taken at least six times, preferably twice per company shift. In case of significant variations, continuous measurements will be made.

2.2.2. Measurements continuous

This type of electrical measurement is recommended if more accurate and reliable results are desired. These measurements are performed with a network analyzer that remains connected to the equipment to be measured for as long as it is necessary to make the measurement. The interval and duration of the measurement is programmed in the measuring equipment, but the most common is to make measurements every 5 minutes during 24 hours. In case If the behavior is not constant, the measurement interval is increased until the conditions representative of the equipment operation are obtained.

2.2.3. Planning of the measurements.

Planning basically refers to considerations about the type of measurements to be made, decisions about where to measure, the duration of the measurement and the equipment to be used, what provisions need to be made, the risks that may arise and how the measurements are to be made.Electrical measurements should be made in representative or typical periods and not when the company is working in periods of low production or when they have an urgent order and is at full capacity. The following aspects should be considered in the planning of electrical measurements:

• Define the type of measurements to be performed.

• Define representative or typical periods.

• Identification of energetically important loads.

• Assign personnel to perform or supervise electrical measurements.

In the event that the company performs its own measurements, the following should be considered:
• Proper selection of measuring equipment,

• Train operators of measuring equipment.

• Establish record formats.

• Define the frequency of taking readings.

2.2.4. Determination of the variables to be monitored

The number of variables or magnitudes to be measured during the development of a monitoring can vary depending on the information and results to be obtained and the capacity or availability of the measurement equipment to be used. It is logical to think that the greater the number of variables to be monitored, the better the evaluation of the equipment and characterization of the use of electrical energy in the facilities under study and, consequently, the estimates made in the energy diagnosis will be better supported and the margin of uncertainty of the expected results will be smaller.The main electrical quantities or variables to be measured include the following:

Table 2 Electrical Variables and Units.

Variable electrical	Unit
Power consumption	kWh
Active power	kW
Reactive power	kVAR
Power factor	%
Current intensity	A
Voltage or potential	V

The above variables are considered indispensable to obtain a picture of the electrical behavior of an equipment, system or installation; measurements of energy consumption (kWh), active power (kW) and reactive power (kVAR) are considered indispensable for the purpose of energy diagnosis, while data on current (A) and voltage (V) are considered indispensable for the purpose of energy diagnosis. provide additional informationOn the performance of the installation and the quality of supply. The power factor case can be obtained from the active power and reactive power values.

2.2.5. Identification of the points of measurement.

The selection of measurement points within a user's installations is one of the most important aspects in the monitoring of electrical variables and is conditioned by a series of factors, among which the following stand out: the complexity of the installation, the availability of measurement equipment, the number of consumption and demand measurements to be obtained throughout the building, and the ease of access for the installation of the equipment.The complexity of the installation will depend on the size of the company, the number and type of loads to be fed, the number of nodes included in the installation and fed from the general supply (substation busbars, load centers, distribution boards). In the case of installations where electrical energy is supplied and distributed internally at medium and high voltage, it is recommended to select the equipment that will have the greatest impact on the level of energy demand and consumption for each of the voltage levels present in the various sections of the installation. In this particular case, it should also be considered the need to have auxiliary equipment for connection of measuring equipment (current transformers and potential transformers) with the characteristics and insulation levels appropriate for the voltage level where they are used as well as the magnitude of the loads to be fed.On the other hand, for low voltage installations, it is generally not required that potential transformers, and current transformers may be required for those loads or circuits that exceed the rated current capacity of the measuring equipment.

2.3. MEASURING INSTRUMENTS AND VARIABLES TO MONITOR

There are different types of continuous measurement equipment, and their selection is based on the characteristics they can offer, as well as the ease of use and acquisition; among the most common are the following:

2.3.1. Ammeter

The ammeter is an instrument that measures the line current in an electrical system in amperes. There is a wide variety of ammeters but the most commonly used for measuring current in installed systems is the so-called hook meter (see Figure 2.1.1). This is a portable instrument that gives a direct reading of the current flowing in the conductor to which it is connected.

Figure 2 Ammeter.

For points where current monitoring is considered important, fixed meters connected in series with the load where the current flow is measured are used. They are currently available with analog or digital readers.

2.3.2. Hook ammeter with power function
Characteristics:
- TRMS measurement

- Accepts conductors up to 26 mm

- Audible continuity.

- Auto shutdown function

- Low battery indication

- Data freezing and spikes

- Safety category CAT III 600 V

Table 3 Characteristics of the Power Hook Ammeter.

AC Amperes	40/400/600 Amperes
AC-DC voltage	600 Volts
Resistance	999.9 Ω
Frequency	5.0 Hz to 500 Hz
Active power (W)	0-360 KW
Reactive Power (VAR)	0-360 KVAR
Apparent Power	0-360 KVA
Power factor	0.10 a 0.99
Feeding	2 AAA batteries
Dimensions	266 x 177 x 81 mm
Weight	224 gr
Included accessories	Test leads Alligator clips Instruction Manual

2.3.3. Clamp ammeter for quality measurement electrical

This clamp meter combines the advantages of a power quality analyzer, a power quality recorder and a clamp meter in a single instrument ideal for monitoring electronic loads.

Figure 3 Hook ammeter with power function.

Applications

Configuration and troubleshooting of variable speed drives and interruptible power supplies: Check proper operation by measuring the key parameters that define power quality.

Harmonic measurements: Identify problems caused by harmonics that may damage or affect critical equipment.

Pickup of inrush currents: Check the inrush current resulting from false restarts or unexpected tripping of circuit breakers.

Load studies: Check the capacity of the electrical installation before adding new loads.

2.3.4. Luxmeter

The luxmeter is a device that measures light intensity in lux. This device is designed to show the relationship between light intensity and angle of incidence.

Figure 5 Luxmeter

The recommended lighting levels for a specific place or area depend on the activities to be carried out there. In general, a distinction can be made between tasks with minimal, normal or demanding lighting requirements.In order to determine the appropriate type of lighting for the site, the following lighting standards should be taken into account:

Table 4 Recommended lighting levels for interior and exterior lighting

ACTIVIDAD	DENSIDAD DE POTENCIAELECTRICA (W/m^2)	
	ALUMBRADO INTERIOR	ALUMBRADO EXTERIOR
Oficinas	16,0	1,8
Escuelas	16,0	1,8
Hospitales	14,5	1,8
Hoteles	18,0	1,8
Restaurantes	15,0	1,8
Comercios	19,0	1,8
Bodegas o áreas de almacenamiento.*	8,0	
Estacionamientos interiores.*	2,0	

NOM-007-ENER-1995 "Energy efficiency for lighting systems in **non-residential buildings.**

Table 5 Recommended illumination levels for work centers

TAREA VISUAL DEL PUESTO DE TRABAJO	ÁREA DE TRABAJO	NIVELES MÍNIMOS DE ILUMINACIÓN (LUX)
En exteriores: distinguir el área de tránsito, desplazarse caminando, vigilancia, movimiento de vehículos.	Áreas generales exteriores: patios y estacionamientos.	20
Requerimiento visual simple: inspección visual, recuento de piezas, trabajo en banco y máquina.	Áreas de servicios al personal: almacenaje rudo, recepción y despacho, casetas de vigilancia, cuartos de compresores y palería.	200
Distinción moderada de detalles: ensamble simple, trabajo medio en banco y máquina, inspección simple, empaque y trabajos de oficina.	Talleres: áreas de empaque y ensamble, aulas y oficinas.	300
Distinción clara de detalles: maquinado y acabados delicados, ensamble e inspección moderadamente difícil, captura y procesamiento de información, manejo de instrumentos y equipo de laboratorio.	Talleres de precisión: salas de cómputo, áreas de dibujo, laboratorios.	500
Alta exactitud en la distinción de detalles: ensamble, proceso e inspección de piezas pequeñas y complejas y acabado con pulidos finos.	Áreas de proceso: ensamble e inspección de piezas complejas y acabados con pulido fino.	1,000
Alto grado de especialización en la distinción de detalles.	Áreas de proceso de gran exactitud.	2,000

NOM-025-STPS-1999 "lighting conditions in workplaces".

2.4 NETWORK ANALYZER

There is on the market an instrument capable of obtaining all the information of the electrical parameters instantaneously, called network analyzer, this type of instrument has a system of acquisition and recording of information, so they have the ability to print the measured data for continuous recording in preset periods as they are programmable and can be used for the following records:

The overall plant load over a day or longer period to determine the demand profile and identify the magnitude and timing of peak demand. To carry out this measurement it is convenient to connect them to the secondary output of the main transformer(s). It is convenient to take into account that normally this equipment is for low voltage measurement, up to 600 V, so in some cases it will be necessary to use voltage transformers (doughnuts).

• The overall overnight load in an area or department to observe the behavior of loads over time

• In general, the start-up of major electrical loads such as power systems, air conditioning, refrigeration, lighting and other electrical energy consuming systems.

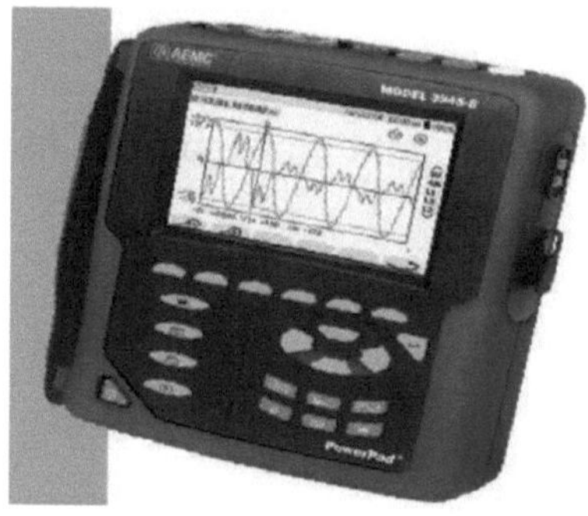

Figure 6 Network analyzer

2.4.1. Graph of current

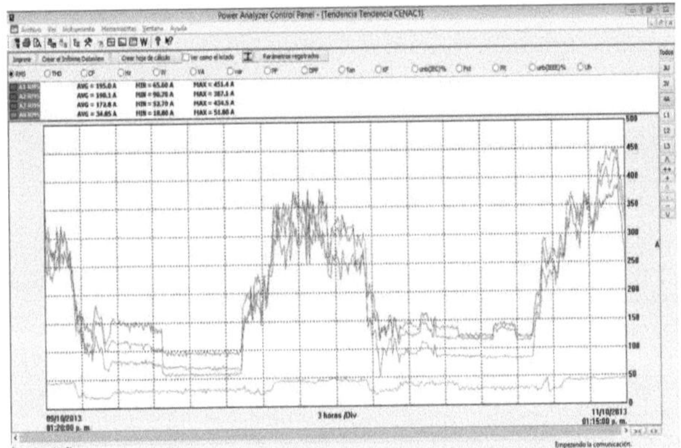

Figure 7 Example of Current Graph

2.4.2. Active power graph

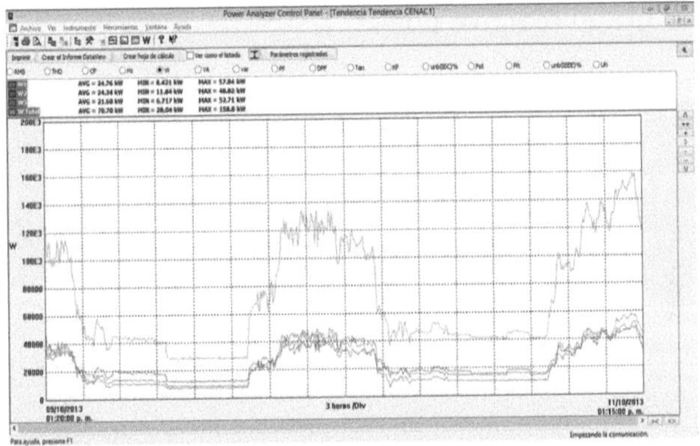

Figure 8 Example of Active Power Graph

2.4.3. Power graph apparent power

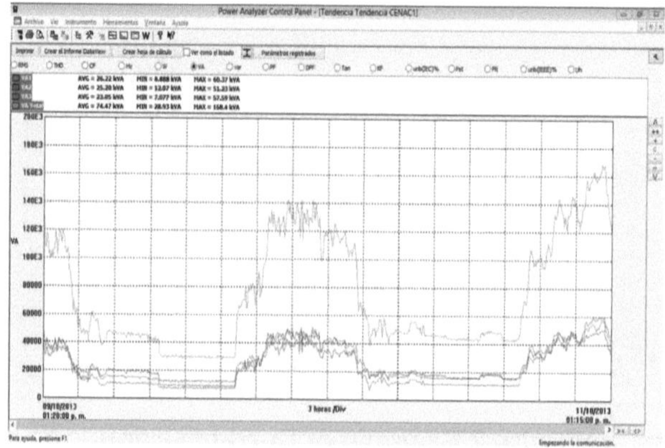

Figure 9 Example of Apparent Power graph.

2.4.4. Power graph reactive power

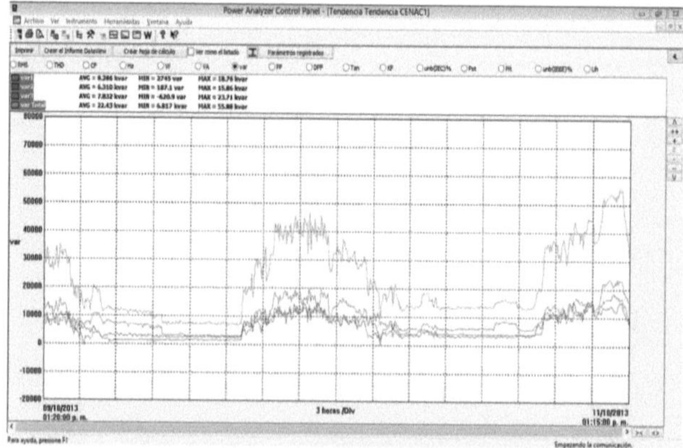

Figure 10 Example of Reactive Power Graph

2.4.5. Graph of harmonics with respect to the current

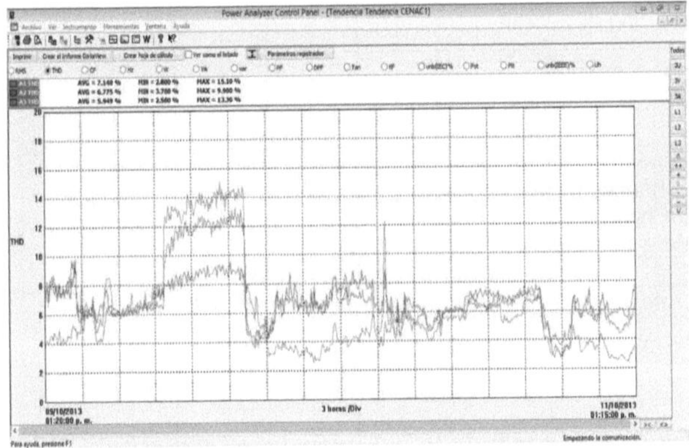

Figure 11 Example of harmonics with respect to current.

2.4.6. Graph of harmonics with respect to voltage

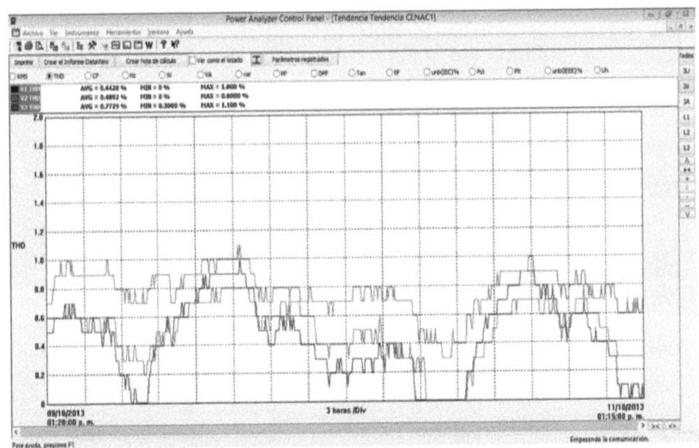

Figure 12 Example of Harmonics with respect to Voltage.

The graphs shown above are only intended to complement the analysis, which is performed with data recorded in the measurement, these are downloaded from the analyzer program on the PC in order to perform the desired calculations. As an example, part of the data to be analyzed is shown. It should be remembered that the analyzer is programmed every 5 minutes to record the time to be analyzed.

Model 3945	Series										
Trend	CENAC1										
Starting day	Start time										
09/10/2013	01:20:00 p. m.										
Type of connection: 3-Phase 4-Wire											
Date	Time	W1	W2	W3	W Total		Wh 1	Wh 2	Wh 3	Wh Total	
		W	W	W	W		Wh	Wh	Wh	Wh	
09/10/2013	01:20:00 p. m.	4616 8.8	3632 3.9	4220 9	124 701 .7		384 7.4	302 6.9 9	351 7.4 2	103 91.8 1	
09/10/2013	01:25:00 p. m.	3716 1.7	3053 1.7	3833 7.9	106 031 .3		694 4.2 1	557 1.3	671 2.2 5	192 27.7 5	8835 .94
09/10/2013	01:30:00 p. m.	3591 0.2	2958 8.8	3798 1.6	103 480 .6		993 6.7 3	803 7.0 3	987 7.3 8	278 51.1 4	8623 .39
09/10/2013	01:35:00 p. m.	3884 4.3	3423 2.9	4017 6.8	113 254		131 73. 75	108 89. 77	132 25. 45	372 88.9 7	9437 .83
09/10/2013	01:40:00 p. m.	4035 6.8	3370 9.2	3780 1.1	111 867 .1		165 36. 82	136 98. 87	163 75. 54	466 11.2 3	9322 .26
09/10/2013	01:45:00 p. m.	3538 2.1	2959 7.9	3376 2.8	987 42. 76		194 85. 32	161 65. 36	191 89. 1	548 39.7 9	8228 .56
09/10/2013	01:50:00 p.m.	3674 6.9	3042 5.9	3381 8.8	100 991 .6		225 47. 56	187 00. 86	220 07. 33	632 55.7 5	8415 .96
09/10/2013	01:55:00 p. m.	3505 2.3	2906 6.6	3226 8.2	963 87. 04		254 68. 59	211 23. 07	246 96. 35	712 88.0 1	8032 .26
09/10/2013	02:00:00 p. m.	3635 7.9	3052 7.8	3477 3.2	101 658 .9		284 98. 41	236 67. 05	275 94. 11	797 59.5 8	8471 .57
09/10/2013	02:05:00 p. m.	3181 3.7	3117 9.8	3377 1.2	967 64. 71		311 49. 55	262 65. 37	304 08. 38	878 23.3 1	8063 .73
09/10/2013	02:10:00 p. m.	3539 0.8	3225 9.5	3539 8.3	103 048 .6		340 98. 79	289 53. 66	333 58. 24	964 10.6 9	8587 .38
09/10/2013	02:15:00 p. m.	3711 6.2	3163 3.1	3473 7.1	103 486 .5		371 91. 8	315 89. 75	362 53	105 034. 56	8623 .87
09/10/2013	02:20:00 p.	4163 63	3446 46	3865 65	114 758		406 61.	344 62.	394 74.	114 597.	9563 .21

	m.	3.1	9	6.4	.5		23	17	37	77	
09/10/2013	02:25:00 p.m.	38090.6	33669.3	35975.2	107735		43835.44	37267.95	42472.3	123575.69	8977.92
09/10/2013	02:30:00 p.m.	34552.3	31692.4	32190.5	98435.19		46714.8	39908.98	45154.84	131778.62	8202.93
09/10/2013	02:35:00 p.m.	38949.7	33724.7	33869.5	106543.9		49960.6	42719.37	47977.3	140657.28	8878.66
09/10/2013	02:40:00 p.m.	41777.2	34842.1	38369.5	114988.8		53442.03	45622.89	51174.76	150239.67	9582.39
09/10/2013	02:45:00 p.m.	39319.2	35278.2	36900.6	111498		56718.63	48562.74	54249.81	159531.18	9291.51
09/10/2013	02:50:00 p.m.	37275.4	32548.3	36366.5	106190.1		59824.91	51275.09	57280.35	168380.35	8849.17
09/10/2013	02:55:00 p.m.	40697.3	35758.4	38635.6	115091.3		63216.36	54254.96	60499.98	177971.3	9590.95
09/10/2013	03:00:00 p.m.	39918.4	34663.1	37231.8	111813.3		66542.9	57143.55	63602.62	187289.07	9317.77
09/10/2013	03:05:00 p.m.	39005.3	33842.9	36710.7	109558.9		69793.34	59963.79	66661.85	196418.98	9129.91
09/10/2013	03:10:00 p.m.	37537.5	33430.4	32232.4	103200.3		72921.46	62749.66	69347.88	205019	8600.02
09/10/2013	03:15:00 p.m.	34566.8	32183.1	32837.2	99587.21		75802.03	65431.59	72084.32	213317.94	8298.94
09/10/2013	03:20:00 p.m.	31641.4	33744.5	35369.7	100755.7		78438.82	68243.63	75031.8	221714.24	8396.3
09/10/2013	03:25:00 p.m.	33028.6	35521	34660.9	103210.5		81191.2	71203.71	77920.21	230315.12	8600.88
09/10/2013	03:30:00 p.m.	27806.2	30631.4	30282.4	88719.96		83508.38	73756.33	80443.74	237708.45	7393.33
09/10/2013	03:35:00 p.m.	27073.8	31742.1	29600.2	88416.06		85764.53	76401.5	82910.42	245076.46	7368.01

Table 6 Example of analysis of data recorded in the analyzer.

2.5 SECURITY IN THE MEASUREMENTS.

It is necessary and very important to observe the following safety
recommendations:
I. Consider that all equipment and installations are energized.
II. Connect the neutral or ground first and then the phases. When
removing the equipment, proceed in reverse order.

III. Make sure that the board and equipment is firmly energized,
otherwise the results will be altered and they will be exposed to serious
dangers.
IV. Clearly identify the operating voltage and check that the measuring
equipment is adequate.
V. Preventing the opening of a current transformer

VI. Electrical measurements must be made on the load side of the
protection equipment in order to protect from a possible short circuit,
avoid measuring potentials on very close points.
VII. Never perform a measurement so quickly that it endangers the
equipment or the physical integrity of persons.
VIII. Use the proper tool and protective equipment.

IX. Avoid testing on the equipment.

X. Avoid making temporary repairs to measuring equipment by
inexperienced personnel.
XI. Verify historical information about the equipment you are measuring.

XII. If the environment is corrosive, check the connections and clean the
parts to be connected.
XIII. Train your collaborators in the application of measures and
medications in case of an electrical accident.

2.6 MEASUREMENT ANALYSIS ELECTRICAL

The analysis of the electrical measurement is done to compare the
electrical parameters with the electrical billing. The electrical parameters
of the general measurement must be similar to the billing; otherwise,
there are two possibilities: the electrical measurement was not correct
due to equipment failures or there are undue electrical charges, which
can be claimed to the utility company. On the other hand, it serves as a

background prior to the implementation of energy saving measures and shows the number of hours of electricity consumption of all electrical equipment in operation in the company. With the electrical measurement it is possible to detect possible unbalance between phases,which cause areduction ofthe lifeof the equipment operating in three phases, premature aging of the conductors. Also The electrical measurement shows the times when the electrical demand peaks occur and their magnitude, for which actions can be implemented to reduce them and, consequently, reduce the amount of electrical billing. The electrical measurement shows the hours where the electrical demand peaks occur and their magnitude, for which actions can be implemented to reduce them and, as a consequence, reduce the amount of the electrical invoicing.In the case of companies, the electrical measurement graph of active power versus production is useful to detect possible anomalies in the use of electrical energy and thus establish whether it is necessary to switch on programs, disconnect equipment during the peak period and identify equipment that can be operated in tariff periods where electrical energy is more economical.

3.1. EXAMPLE OF ENERGY DIAGNOSIS

3.1.1. Reason social.
CAMP SPORTSWEAR SR L DE CV

3.1.2. Name commercial.
CAMPECHE SPORTSWEAR

3.1.3. Address
Campeche-Hampolol Highway km 4.5, Fidel Velázquez, C.P. 24023. Campeche, Campeche

3.2. SERVICE DATA

Table 7 Main data identifying the service.

RPU	789030805651.
	NOMBRECAMP SPORTSWEAR SR L DE CV
ACCOUNT	81DW04B238121600
RATE	HM
TYPE OF SUPPLY	High-High
INSTALLED LOAD	498
CONTRACTED DEMAND	498
METER	0X3X35
MULTIPLIER	350

All readings taken fromthe meter when checking your consumption and demand must be multiplied by the multiplier to obtain the actual value.

3.3. SUMMARY OF TARIFA

Table 8 Medium Voltage hourly rate.

TARIFA	ELEMENTOS DE LA FACTURACIÓN							
	CONSUMO KWh			DEMANDA FACTURABLE KW	TIPO DE SUMINSITRO	FACTOR DE POTENCIA	DAP	IVA
HM	BASE	INTER	PUNTA	DEMANDA FACTURABLE	MEDIA TENSION CARGO	√	√	√

Table 3.3 shows the elements that make up the billing in which this service is contracted.

Figure 13 Hourly Rate Periods

DEMAND: Power that an equipment needs to turn on.

CONSUMPTION: It is the power needed to turn on the equipment for the time it is used.

Table 9 Amounts for the month of March.

AMOUNTS FOR THE MONTH OF MARCH

	CONSUMP	DEMAND
KWH BASE $191.60	$0.	7706KW
KHW	$0.9357	
KWH POINT	$1.8858	

In order to implement demand management, it is necessary to know the periods and times when special attention should be paid to identify equipment that can be turned off or processes that can be rescheduled so as not to affect monthly billing.

3.4. DATA HISTORICAL DATA.

3.4.1. Cost history average cost

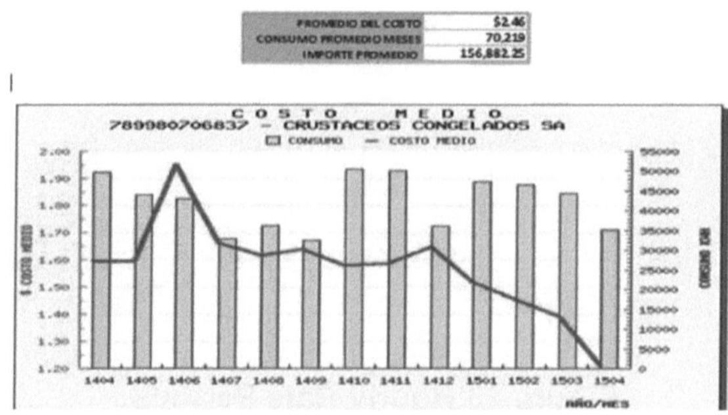

Figure 14 Average cost

The graph shows the Average Cost (**price we pay for each kWh of energy**). This is determined from the total of what we pay (**$**), between what we consume (**kWh**).

3.4.2. Consumption history in the different periods.

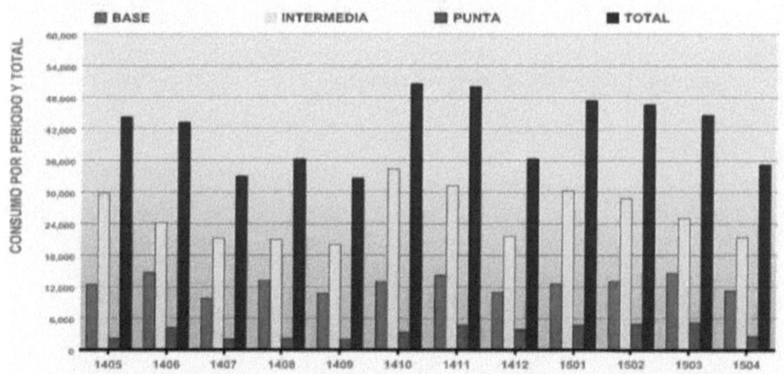

Figure 15 Consumption History by Period.

It is observed that the highest energy consumption is concentrated in the intermediate hours, this behavior is due to the fact that this period comprises the most hours.

FECHA	BASE	INTER	PUNTA	FACTURABLE
201404	68	150	67	92
201405	117	160	62	92
201406	137	167	165	166
201407	131	73	65	77
201408	71	71	68	69
201409	51	67	49	55
201410	77	179	68	99
201411	103	162	65	95
201412	66	135	62	84
201501	66	156	69	96
201502	77	159	62	92
201503	67	162	63	93
201504	57	64	57	60

Figure 16 Consumption History by Period.

The peak period is where care must be taken, as the KWH Price increases at twice the rate of the base and intermediate period.

3.4.3. Demand history by period and billable demand .

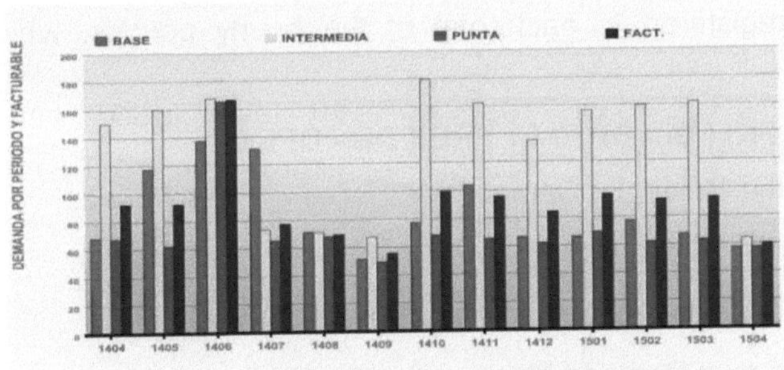

Figure 17 Demand History by Period.

It is observed that the higher the peak demand (PD), the higher the Billable Demand (Demand charged by CFE), since it takes 100% of this value.

3.4.4. Demand history billable.

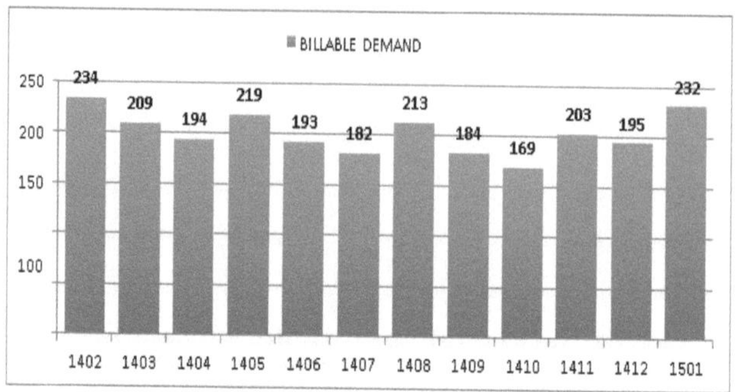

Figure 18 Billable Demand History.

The Billable Demand is used to establish the Electric Energy bill. The maximum Billable Demand recorded in the history is **234 KW**.The Billable Demand formula is integrated by means of the maximum demands registered in each one of the hourly periods, which is as follows:

DF = DP + FRI (DI - DP, 0) + FRB (DB - DPI, 0)

✓ PD: It is the Maximum Demand Measured in Peak Period.

✓ DI: Maximum Demand Measured in Intermediate Period.

✓ DB: Is the Maximum Demand Measured in Base Period.

✓ DPI: Maximum Demand Measured in the Peak and Intermediate Periods.

✓ FRI and FRB: Reduction Factors, depending on the Region and Tariff. FRI=0.3, FRB=0.15

In simpler terms, the formula states that 100% of the peak demand will be charged, plus the application of a reduction factor of 30% for the difference of the peak demand minus the base and a reduction factor of 15% for the difference of the base demand minus the greater of the peak or the intermediate.As mentioned above, it is possible to achieve billing reductions by applying savings measures during peak hours.

Demand Control and Management: It is defined as the **action of interrupting for time intervals the operation of certain electric loads that directly affect the Billable Demand**.

To carry out a Demand Control and Management is to identify which of the equipment available can stop operating during peak hours or can exchange their operating time to another period; this is due to the fact that these hours represent a considerable percentage of the billing. It should be taken into account that in order to implement the Demand Management in the If the Peak Demand is the highest this value will be equal to the Billable Demand. If the Peak Demand is the highest this value will be equal to the Billable Demand.

3.5. MEASUREMENT AND ANALYSIS OF PARAMETERS.

In its billing history, it can be seen that it is possible to implement Demand Control and Management during the **Peak Period**.

MES	PUNTA	INTERMEDIO	BASE	BILLABLE DEMAND
Feb-14	223	259	244	234
Mar-14	207	215	182	210
Apr-14	175	238	201	194
May-14	207	246	212	219
Jun-14	179	226	185	194
Jul-14	170	210	175	182
Aug-14	202	238	224	213
Sep-14	170	216	183	184
Oct-14	168	172	173	170
Nov-14	192	226	233	204
Dec-14	188	209	215	196
Jan-15	220	260	225	232
PROM	192	227	205	203
MAX	223	260	244	234
MIN	168	172	173	170

Figure 19 Billing history.

The highest demand recorded during this peak period was **223 KW**, and the lowest was **168 KW**.

3.5.1. Analysis of consumption.

The following consumptions were obtained with the network analyzer:

CYCLES PER DAY					
CONSUMPTION PER HOURLY PERIOD					
DATE	TIME	BASE KWH	INTERMEDIATE KWH	KWH POINT	TOTAL KWH
16/02/2015	4:00:00 p.m. 11:55:00 p.m.	0.00	517	510	1,027
17/02/2015	12:00:00 a.m. 11:55:00 p.m.	443	1,309	536	2,288
18/02/2015	12:00:00 a.m. 11:55:00 p.m.	398	1,912	430	2,739
19/02/2013	12:00:00 a.m. 3:35:00 p.m.	450	1,351	0	1,801
TOTALS		1,290	5,089	1,476	7,856

Figure 20 Consumption analysis by hourly period.

It can be observed that during the peak period there is a change of habit, since consumption is similar to that of the base period, showing that equipment is left on unnecessarily. However, there is a decrease in peak consumption from 536 to 430 KWH.

3.5.2. Analysis of demand.

Point measurements were made with the AEMC 3945-B Network Analyzer, where the measured demand curves are shown below.

CYCLES PER DAY					
MAXIMUM AND BILLABLE DEMAND					
DATE	TIME	BASE KW	INTERME DIATE KW	PUNTA KW	DEMAND BILLABLE
16/02/201 5	4:00:00 p.m. 11:55:00 p.m.	0	199	189	192
17/02/201 5	12:00:00 a.m. 11:55:00 p.m.	149	209	210	210
18/02/201 5	12:00:00 a.m. 11:55:00 p.m.	141	233	188	201
19/02/201 3	12:00:00 a.m. 3:35:00 p.m.	136	217	0	65
MAXIMUM		149	233	210	217

Figure 21 Peak and billable demand analysis

The analysis shows that demand management is possible, since equipment can be turned off during the peak period.

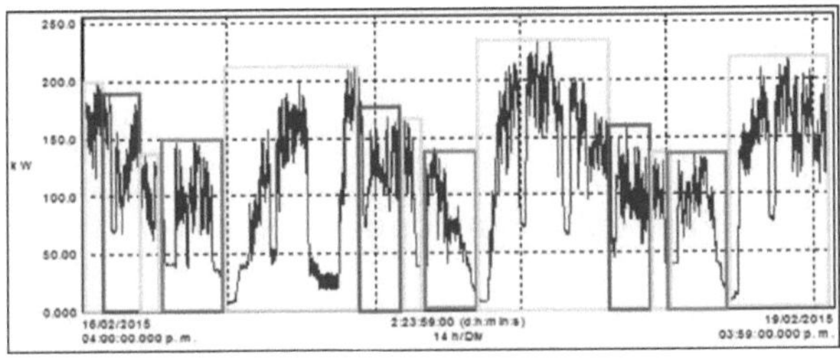

Figure 22 Comparative graph of the base, intermediate and peak periods, extracted from the network analyzer.

45

It can also be noted that from 5:30 am to 7:00 am (base time) there is a demand of 15 KW during the three days of the analysis, it is observed that the maximum demand of 149 KW in the base time occurs from 12:22 am to 5:30 am during the three days.

3.5.3. Analysis of transformer.

The measurements on this transformer were performed on February 16, 2015 from 4:00 pm. Until February 19 of the same year until 3:35 pm. With the power analyzer AEMC 3945-B, this transformer is located in the facilities of the maquiladora CAMP SPORTSWEAR S R L D E CV. The following graphs show the behavior of the substation. It is operating at 33% of its capacity.

Nombre	PROM	MIN	MAX	Unidades
VA Linea1	38.994k	0.0000	85.166k	VA
VA Linea2	43.181k	0.0000	86.944k	VA
VA Linea3	38.493k	0.0000	79.332k	VA
VA Suma de Fases	120.67k	0.0000	245.80k	VA

Figure 23 Apparent power demand.

According to the above measurement and the transformer capacity of **750 KVA**, an apparent power demand of **245.80 KVA** is recorded.

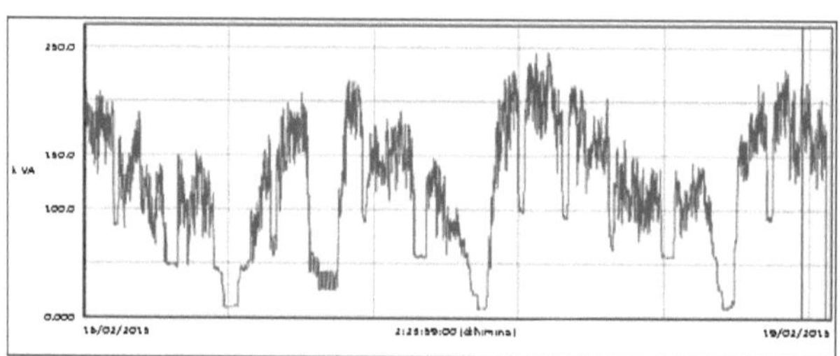

Figure 24 Transformer kva curve

3.5.4. Analysis of current

Nombre		MIN	MAX	Unidades
Arms Linea1	326.98	0.0000	639.30	A
Arms Linea2	358.78	0.0000	653.30	A
Arms Linea3	316.51	0.0000	593.80	A

Figure 25 Current analysis of each line

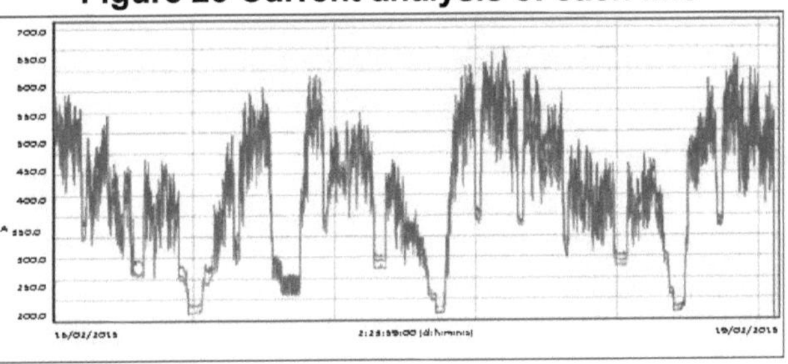

Figure 26 Phase and neutral current curves.

The figure shows the current curves in phases F1, F2, F3 and NEUTRAL of the transformer on the Y axis, in addition to the curves of the maximum and minimum value in relation to the rated current, while on the X axis the period in days, where it is considered that from +-5% of current variation can be detrimental to the operation of the equipment. In this graph we can see that the current variation within the allowable range having an average current from phase to phase of 9.04%.

3.5.5. Analysis of harmonics

Harmonics are distortions of voltage and/or current sine waves in electrical systems.The equipment installed in places where these exist, do not work in nominal conditions, since neither the voltage nor the current will be those marked on the nameplate. Therefore, equipment installed in places or plants where harmonics exist will not operate properly.

Nombre	PROM	MIN	MAX	Unidades
Athd Linea1	7.4639	0.0000	34.100	%
Athd Linea2	8.7106	0.0000	41.300	%
Athd Linea3	10.198	0.0000	55.000	%

Figure 27 Current analysis of harmonics in each phase.

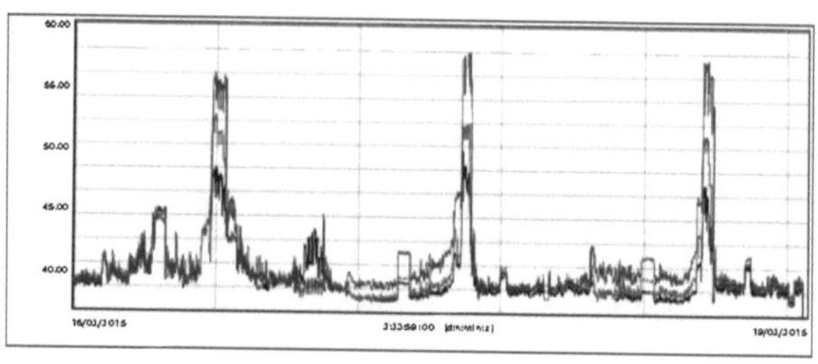

Figure 28 Current distortion curves.

In the behavior of the currents, it was detected that an average distortion degree of 43.47% was reached, which means that it is out of the permissible range according to the CFE L0000-45 specifications and the IEEE519 standard, since it exceeds the 10% distortion allowed in the electrical system.Analyzing the behavior of the voltage in its installations, it was detected that the maximum harmonic distortion percentage is 1.9%, which means that it is within the allowable range according to the CFE L0000-45 specification and the IEEE 519 standard, which is 5% (see figure 3.5.5.3).

Nombre	PROM	MIN	MAX	Unidades
Vthd Linea1	0.733	0.0000	1.1000	%
Vthd Linea2	0.650	0.0000	1.1000	%
Vthd Linea3	0.626	0.0000	1.1000	%

Figure 29 Voltage analysis of harmonics in each phase.

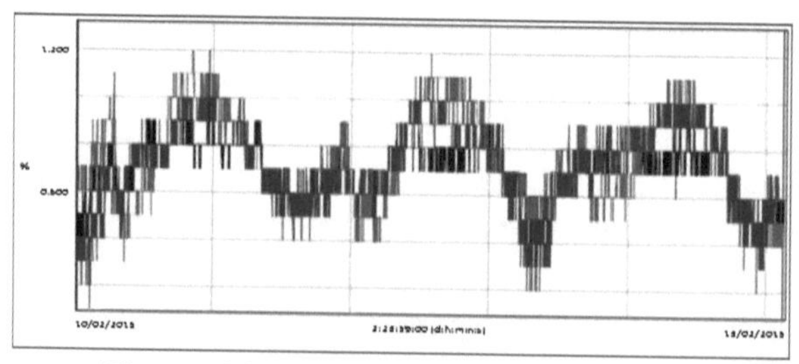

Figure 30 Voltage distortion curves.

3.5.6. Power factor analysis (correction)

It is an indicator of the correct use of energy; in general, it is the amount of energy that has been converted into work.The indicators that CFE manages for the degree of Energy Efficiency:
- Power factor greater than 90% = Bonus (Up to 2.5% of billing).

- Power factor less than 90% = Surcharge (Up to 120% of billing).

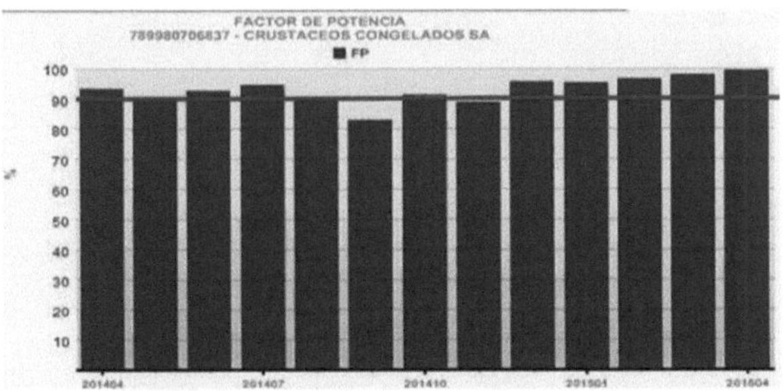

Figure 31 Power factor history

An adequate power factor should be =/> than 90%. Currently the Average Power Factor of a Year's History is 87.24%, which means that the energy it receives is being converted into useful work. Similarly, in its billing there is an average surcharge of approximately 1.9 %, which means that the energy it receives is being converted into useful work. whichrepresents$2,447.00 per month approximatelyHowever, in one year of penalty a total amount of **$ 29,367.07** has been paid.

FECHA	CARGO	%
1402	$600.70	0.44%
1403	$2,818.28	2.19%
1404	$2,280.83	1.71%
1405	$3,412.31	2.46%
1406	$4,520.94	3.55%
1407	$4,234.42	3.45%
1408	$2,516.16	1.64%
1409	$3,569.18	2.83%
1410	$1,856.99	1.66%
1411	$1,499.09	1.25%
1412	$1,898.66	1.67%
1501	$159.53	0.12%

Figure 32 Low power factor penalty history

3.5.6.1. Power factor correction.

Nombre	PROM	MIN	MAX
PF Linea1	0.872	0.0000	0.982
PF Linea2	0.872	0.0000	0.973
PF Linea3	0.865	0.0000	0.979

Figure 33 Analysis of the average power factor of each line.

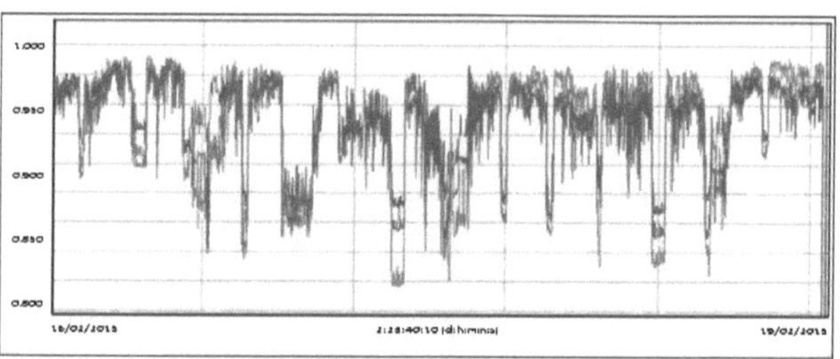

Figure 34 Power factor curves

This figure shows the power factor curves of phases F1, F2 and F3, as well as the power factor (PF) of the minimum (Min) and maximum (Max) values. The X-axis shows the period in days, while the Y-axis shows the values of the index established by "the regulator", the results of the curves show the power factor, which are not in the ranges established by the regulator, where it can be seen that the average P.F. is 87.24 %.

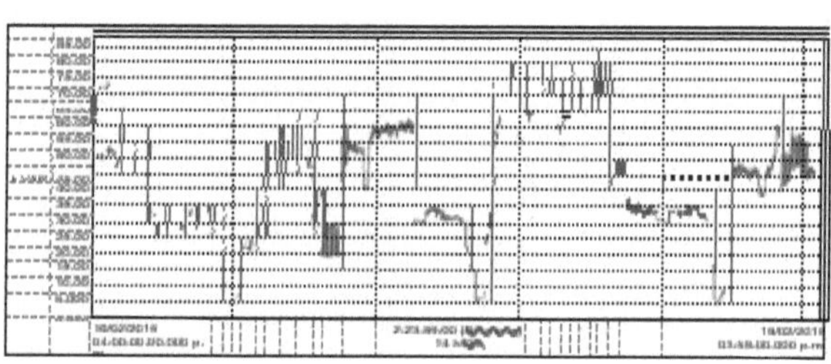

Figure 35 KVARS Curve in the System

Power factor correction (calculation).

Delayed power factor = .87

$$f.p.\ lagged = \cos 0 = \cos^{-1}(.87) = 29.54°.$$

Apparent power S_1

Data:

$P = 498\ KW = 498000\ W$ installed load

$cp_1 = 29.54°$ phase angle between voltage and current

$$S_1 = \frac{P}{\cos_{cp1}} = \frac{498000\ W}{\cos(29.54°)} = 572406.0904\ VA$$

Reactive power Q_1

$$Q_1 = \sqrt{S_1^2 - P^2} = \sqrt{(572406.0904VA)^2 - (498000W)^2} = 282213.9832\ VAR$$

It is desired to achieve a power factor around 95% with respect to the 87% that currently prevails.

Desired power factor = .95

$$f.p. = \cos 0 = \cos^{-1}(.95) = 18.19°.$$

Desired apparent power S_2

Data:

$P = 498\ KW = 498000\ W$ installed load

$cp_2 = 18.19°$ phase angle between voltage and current

$$S_2 = \frac{P}{\cos_{cp2}} = \frac{498000\ W}{\cos(18.19°)} = 524195.8765\ VA$$

Desired reactive power Q_2

$$Q_2 = \sqrt{S_2^2 - P^2} = \sqrt{(524195.8765VA)^2 - (498000W)^2} = 163637.7615\ VAR$$

The capacitor bank to be installed should supply three-phase reactive power

$$Q = \sqrt{3}\ V \cdot I = Q_1 - Q_2 = 282213.9832 - 163637.7615 = 118576.2217\ VAR$$

$$Q = 118576.2217\ VAR = 118.576\ KVAR$$

According to commercial capacitor banks, a 120 KVAR capacitor is recommended.

3.6. RECOMMENDATIONS.

3.6.1. Control and management of the demand

In order to control and manage the demand, it is **recommended to turn off the Air Conditioners 20 minutes before and 20 minutes after the peak period according to the schedule (Summer - Winter)**, this is possible because during the **Base Period** it can be noticed that not all the Air Conditioners are turned on, only on weekends only what is necessary.

Día de la semana	Base	Intermedio	Punta
lunes a viernes	0:00 - 6:00	6:00 - 20:00 22:00 - 24:00	20:00 - 22:00
sábado	0:00 - 7:00	7:00 - 24:00	
domingo y festivo	0:00 - 19:00	19:00 - 24:00	

Figure 36 Daylight saving time

Día de la semana	Base	Intermedio	Punta
lunes a viernes	0:00 - 6:00	6:00 - 18:00 22:00 - 24:00	18:00 - 22:00
sábado	0:00 - 8:00	8:00 - 19:00 21:00 - 24:00	19:00 - 21:00
domingo y festivo	0:00 - 18:00	18:00 - 24:00	

Figure 37 Winter timetable

3.6.2. Demand simulation billable

By performing a Peak Period Demand Control and Management by reducing 20% of its peak demand of 220 KW (January 2015) the following savings can be achieved.

CURRENT BILLABLE DEMAND			
DP =	220	Region: Peninsular	
DI =	260	FRI:	0.3
DB =	225	FRB:	0.15
DF = DP + FRI max(DI - DP,0) + FRB max(DB - DPI,0 DF=220+0.3(260-220)+0.15(225-260)=232KW $ 43,319.04			

Figure 38 Current billable demand.

PROPOSED BILLABLE DEMAND			
DP =	176	Region: Peninsular	
DI =	260	FRI: 0.3	
DB =	225	FRB: 0.15	
DF = DP + FRI max(DI - DP,0) + FRB max(DB - DPI,0 DF=176+0.3(260-176)+0.15(225-260)=202KW $ 37,717.44			

Figure 39 Simulation of billable demand.

It is observed that by decreasing the Peak Demand from 220 KW to176,the Billable Demand can be decreased obtaining the following savings.

Ahorros		
$DF	$	186.72
Mensual	$	5,601.60
Anual	$	67,219.20

Figure 40 Savings in billable demand.

We can see that by simply making a change of habit at work we can obtain savings without having to invest.

3.6.3. Capacitor bank power factor (capacity location).

One of the Recommendations to avoid penalties in your Billing, is to Increase your Power Factor although it is recommended to 100%, to obtain 2.5% bonuses in your billing it must be over 90%. By installing a Capacitor bank of **120 KVAR** it will be corrected to 95%.The approximate savings that would be obtained by increasing the power factor to 95% are shown below, which would be your approximate one-year savings.

	AHORROS TOTALES		
	BONIF	CARGOS	TOTAL
MENSUAL	$2,974.04	$2,486.74	$5,460.78
ANUAL	$35,688.46	$29,840.92	$65,529.38

Figure 41 Monthly and annual savings and power factor improvement.

Note: The data presented are estimates; they may vary depending on the importance given to the change of habit in your installations and/or replacement of equipment.

3.6.3.1 Location of capacitors.

The capacitor bank can be placed in different locations,

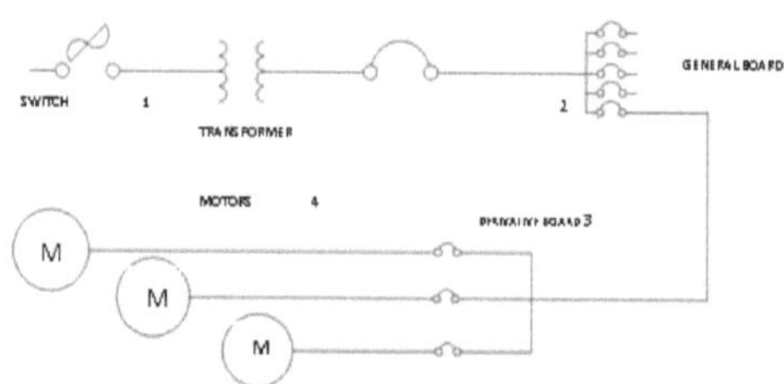

Figure 42 Single-line diagram capacitor bank location.

In position 4 the p.f. is compensated precisely at the place where the reactive power is required. In this way the conductor cross-section of the rest of the installation can be calculated without considering the reactive current consumed by the motor, since it is being generated in the same place where it is consumed. However, the installation cost is high if it is decided to place a capacitor bank for each motor in an electrical installation where there are several motors. Position 3 is practically the same as position 4.In the remaining positions the placement of the capacitor bank is either on the high or low voltage side of the transformer. If the position with the number 1 is chosen, the transformer specification has to include the sum of the active and reactive components of the current. While in the position 2 only the active current of the motor flows through it. The latter solution is also a way to increase the installed capacity of transformers through which a large inductive current flows.

3.6.4. Lighting

o Change T-12 lamps to T5

o Turn off unused building lights.

o Installation of programmers, or presence sensors to reduce usage time.

o Lighting levels allow a low level of illumination to be maintained when a high light intensity is not required.

o Use exterior lights equipped with photocells or timers, which turn themselves off during the day. And with presence sensors for surveillance lights.

o Keeping lamps clean and in good condition could mean a saving of up to 20% in electricity consumption for lighting, since dirty or in poor condition they can lose up to 50% of their luminosity.

3.6.5. Air conditioning.

o It is recommended to clean the filters of your equipment every 30 days.

o Manage the temperature of your equipment to 24 degrees which is a comfortable temperature at which you can be without the use of blankets and at which the equipment will take its breaks properly.

o Keep doors and windows tightly closed to prevent heat from entering the room.

3.6.6. Cooling

○ It is recommended to use timers for water dispensers and refrigerators to program them to stay on during office hours.
○ Install your refrigerator away from heat sources.

○ Do not introduce hot food.

○ The correct position of the thermostat is at the middle level.

○ Unplug the refrigerator and wipe off with a damp cloth any grime that accumulates on the back of the refrigerator at least every two months.

○ Clean the condenser tubes located at the back or bottom of the unit at least twice a year.

CONCLUSIONS OF THE DIAGNOSIS ENERGY BUSINESS.

The average P.F. is **87%,** this means that the current power factor is out of CFE specifications, in other words, this power factor generates surcharges in your billing. The most advisable thing to do is to maintain a power factor above 90% since this will avoid surcharges.To correct this power factor, it is recommended that a 120 KVAR capacitor bank with harmonic filters be installed in the transformer to compensate the power factor and make the electrical system more efficient.It should be noted that capacitor banks help to correct the low power factor, but increase the harmonics in the system, therefore the harmonic filters will help to prevent harmonics from entering your electrical system and not cause any inconvenience in your installations.With the implementation of Change of Habit it is not necessary to invest to obtain savings, this is obtained by implementing changes in the work schedule.In order to have an adequate demand control it is necessary to consume the necessary energy, in the scheme that represents the lowest cost, without affecting the safety, production volume or quality of the product or service.A balance must be found between what is consumed and what is demanded, taking into consideration the schedules for the HM tariff, as in this case.The optimization of processes is essential for an improvement in demand control, it has been detected that there are processes that require minimum amount of energy or even better that do not require electricity, in the tours these processes are those that should be identified in order to reorganize the production line or the operation of the building to obtain benefits and economic savings.There are several methods to control demand, which help to reduce costs.

• Production planning.

• Installation of capacitor bank 120 KVAR

• Installation of Timers, in equipment that are regulated by schedules and thus to be able to avoid that they work of more in not established time.
• PLC installation.

• Installation of sensors, one of the most appropriate are those of temperature, these being present in the air conditioning equipment can be programmed for proper control of the temperature to be set, it has been established that the human body feels in comfort at 24 ° C .

MEDIDAS DE AHORRO DE ENERGIA (MAE)				
Concepto	Ahorro (kWh/Men)	Importe mensual ($)	Ahorro (kW)	Importe Anual ($)
Cambio de Habito	646	$1,299	-	$7,793.07
control de la demanda	-	$5,601.60	30	$67,219.20
Factor de Potencia	-	$4,914.53	-	$58,974.36
TOTAL	646	$11,814.97	30	$133,986.63

Figure 43 Comparative estimated energy savings

BIBLIOGRAPHY

1.- Guide to Support the Development of Energy Diagnostics for Higher Education Institutions (HEI)Serway, R. (2001). Physics, Volume II. (4th Ed.)Pearson Education. 2. Purcell, E. M., Morin D. J. (2013) Electricity and Magnetism. (3rd Ed.) Cambridge University Press.

3. Sears, Z., Young and Freedman (2009). College Physics Vol.2 (12th. Ed.). Pearson Education.

4. Giancoli , D.C. (2008) Física1 Vol.2, (4ª.Ed.). Pearson Educación.
5. Resnick , H. and Krane (2004) Physics Vol.2, (5th Ed.). CECSA.
6. Cabral R., L.G. and Guerrero, R., Virtual Laboratory of Electricity and Magnetism, CIIDET.

7. Walter F.(2012). Physics Java Applets: http://www.walter-fendt.de/ph14s/
8. Franco, A., Physics with Computer, http://www.sc.ehu.es/sbweb/fisica/default.htm
9. Plonus M. A. (1994). Applied electromagnetism. Reverte S. A.
10. Fishbane, P. M., Gasiorowicz S. and Thornton S.T. (1994) Physics for science and engineering. PrenticeHall Hispanoamericana.

Printed by Books on Demand GmbH, Norderstedt / Germany